Natu
Salamar.

Johnston and Schad

Alpha Editions

This edition published in 2022

ISBN: 9789356707313

Design and Setting By

Alpha Editions

www.alphaedis.com

Email - info@alphaedis.com

Contents

NATURAL HISTORY OF THE SALAMANDER, ANEIDES HARDII

The Sacramento Mountains Salamander, *Aneides hardii* (Taylor), is a plethodontid of relict distribution in the spruce-fir vegetational formation from 8500 to 9600 feet elevation in Otero and Lincoln counties, New Mexico. The salamanders on which most of this report is based were collected three, four, and six miles northeast of Cloudcroft in the Sacramento Mountains. Additional individuals were collected on the eastern slope of Sierra Blanca, 1.5 miles southwest of Monjeau Lookout, at about 9000 feet, Lincoln County, and in the vicinity of Summit Springs and Koprian Springs, 9300 feet, Capitan Mountains, Lincoln County. Certain details concerning the populations in Lincoln County will be reported elsewhere (Schad, Stewart, and Harrington, Canadian Jour. Zool., in press).

We would like to thank Mmes. Donna Schad and Lora Lee Johnston, Messrs. Robert Stewart, Frederick Harrington and Ralph Raitt, and Dr. Robert Selander for assistance in the field, Dr. W. Frank Blair and Dr. Marlowe Anderson for the use of specimens in their care, and Dr. A. Byron Leonard for the identification of the molluscan food items.

In the summer rainy season *A. hardii* lives in and under downed timber and under talus accumulations. Occurrence, however, seems to be partly subterranean and always local; seemingly good habitat frequently appears to lack the animals. Our observations and collections were made in July, August, and September in 1956, 1957, and 1958. Two hundred seventy-seven individuals were taken; these were measured, sexed and examined for breeding status. The food and parasite content of the guts of a few individuals was determined.

Thirteen salamanders were kept for varying lengths of time in captivity. The specimens are now stored in collections at New Mexico State University, University of Texas, Museum of Vertebrate Zoology, and the Museum of Natural History, University of Kansas.

The primary study and collecting sites were four and six miles northeast of Cloudcroft, Otero County, at 8600 to 8800 feet in elevation. Vegetation was either almost pure stands of Englemann spruce (*Picea englemanni*) or mixed stands of spruce, Douglas fir (*Pseudotsuga taxifolia*) and white fir (*Abies* sp.). At each locality small oaks (*Quercus*) were present among the dominant conifers. Most of the salamanders found were in downed Douglas fir logs; some were taken from spruce and others from cracks in a variety of deadwood. In the less deteriorated logs the salamanders lived under the loose bark or in small cracks and chambers near the inner bark surface. In large fir logs in advanced stages of decomposition, salamanders could be found to the very centers. This kind of log was apparently highly favorable for salamanders, for it was in such sites that we found notably large numbers of the animals and most of the clutches of eggs that we collected; this kind of log is not frequently found, for its wood is saturated with water and completely punky and nearly ready for final collapse.

In winter, salamanders that spent the summer at the surface presumably move to subterranean cavities, or, at least, to sites away from winter freezing. In December, 1957, and April, 1958, four feet of snow covered our collecting sites, and the downed logs contained ice. A few logs were wet at the surfaces where sunlight hit them, but just under such melt they were icy. On May 3, 1958, snow was in isolated drifts and the centers of the logs were still icy. On May 31, and June 22, 1958, there was no ice anywhere, but no salamanders were evident. Late June is, however, around the earliest time that *A. hardii* emerges (Taylor, 1941).

FOOD AND FORAGING BEHAVIOR

We identified the contents of stomachs from 16 salamanders collected in 1956 and 1957; the items found in them are listed in Table 1. It is not likely that this list is complete for prey species because *A. hardii* eats a variety of food and probably takes prey almost indiscriminately if it is of appropriate size. The kind of food most frequently eaten was ants; they comprised almost 40 per cent of the total items. Nevertheless, less than half the stomachs contained ants; this may mean that salamanders do not make an effort to take ants over any other prey. Such foraging behavior would result in random capture of ants, and it is noteworthy that the frequency distribution of ants in stomachs suggests a Poisson distribution, a mathematical description of one kind of random distribution.

TABLE 1.—NUMBERS OF FOOD ITEMS FOUND IN STOMACHS OF 16 SPECIMENS OF ANEIDES HARDII

ITEMS	Individual animals	Percentage of total (154) individuals	Number of stomachs in which found
Mollusca			
Pupilla muscorum	} 3	1.9%	
Gastrocapta sp.		}	4
Vallonia pulchella?	4	2.5	

Arthropoda			
Arachnoida			
Arachnida	15	9.7	9
Acarina	13	8.4	3
Insecta			
Orthoptera (*Ceuthophilus*)	2	1.3	2
Hemiptera	1	0.6	1
Coleoptera			
adults (carabids and buprestids)	8 }	30.9	7
larvae	38		
Hymenoptera			
ants	62	40.2	7
wasps	2	1.3	2
Unidentified	5	3.2	5

Total	154	100.0	

Adult and larval beetles comprised about 28 per cent of the total items, but were found in only seven of the stomachs. Beetles eaten were small representatives of beetle groups likely to occur in or under logs. A relatively large species of spider was found in nine stomachs; it represented only ten per cent of the items taken but was one of the most important foods when mass is considered.

Two adult salamanders not included in Table 1 were found, in the course of examination for parasites, to have empty stomachs. One was a male, and the other was a female taken from a chamber that held an egg cluster. It would not be surprising regularly to find stomachs empty in "incubating" females, but the fact is that the one other such female collected by us had a small amount of food in the gut; probably these individuals take anything that enters the egg chamber, but do not leave for active pursuit of food.

Foraging behavior of captive salamanders was observed by one of us. The salamanders were maintained in a seven-gallon aquarium, the floor of which was covered with soil, mosses, liverworts, certain flowering plants, and pieces of rotten fir log. The salamanders were placed in the terrarium in September, 1956, July, 1957, and October, 1958; one individual lived 13 months, another 14 months.

A variety of natural foods was present in the soil and plant matter placed in the terrarium, and these were presumably eaten as found by the salamanders. However, the great bulk of the food used by the salamanders was introduced for them, in the form of colonies of *Drosophila melanogaster* in half-pint milk bottles. We tried to keep thriving colonies of flies, primarily of the mutant vestigial-winged type, present in the terrarium; in 1957 this was successful to the extent that there appeared to be a surplus of food available at all times. We did not attempt to feed the salamanders any wholly artificial food, such as ground beef.

Initially, the salamanders, although seemingly healthy and well-fed, were not fat. Those that we maintained on a presumably minimal diet remained slender and did not grow in length. Two individuals captured in 1957, however, were maintained on food in excess, and these grew in length and in girth; from an initial size of about 37 mm. snout-vent length (a subadult size) they attained about 45 mm. snout-vent length (an adult size) in a period of five months. The observations on foraging behavior were made primarily on these latter individuals.

The salamanders captured prey by pursuit. A salamander would pursue a fly until it was caught, or until it moved out of the field of action. The salamanders were attracted by movements of flies, and ignored those that were completely quiet; predation was oriented almost wholly on a visual basis. Once they were within 2 to 4 mm. of a fly they would snap out the tongue to secure the fly; they were successful in capturing vestigial-winged flies in about 75 per cent of all tries. The relative success of capture was greater when the animals were fresh from the field and less after they had become fattened. The vigor of their pursuit also decreased noticeably once they became fat. About two days after any new fly colony was placed in the terrarium, a salamander would take up a position just inside the lip of the milk bottle, which was placed on its side. From this vantage point the salamanders took heavy toll of the fly populations, eating both adults and larvae.

Initially the salamanders foraged indiscriminately in daylight or in darkness. Later, as they became fat, they avoided high light intensity and were active only at night or under artificial light of low intensity. The latter pattern of activity is probably typical of the pattern they maintain under natural conditions. Certainly we never saw individuals abroad in daylight at Cloudcroft, yet under favorable environmental conditions they were to be found in sites that required considerable movement over open areas of the ground surface.

For several months two individuals of *Eurycea longicauda* were kept in with *A. hardii*. Foraging of these two plethodontids is nearly identical, but the tongue of an adult *Eurycea* can be extended somewhat more than one-half inch in capturing flies; for *A. hardii* this distance is usually less than one-quarter inch. The relatively short tongue of *A. hardii* can be correlated with its life in restricted, subsurface chambers, where prey most frequently is close to salamanders; *E. longicauda* inhabits significantly more open sites.

PARASITES

Thirty of the adult *Aneides* collected were examined for parasites; most were parasitized by two species of nematodes, *Oswaldocruzia* sp. and *Thelandros* sp. The former is found in the anterior part of the small intestine and occasionally in the stomach, and the latter occurs in the rectum. There were no gross intestinal pathological changes in the salamanders resulting from parasitism. In fact, no pathological or structural abnormalities were noted in any of the salamanders examined. We believe the two nematodes are well-tolerated by the salamander.

TABLE 2.—OCCURRENCE OF PARASITIC NEMATODES IN ANEIDES HARDII

	Per cent of salamanders infected	Number of nematodes per host		Per cent of nematodes that were immature	
		range	mean	July	Aug.-Sept.
Oswaldocruzia sp.	83	2-15	3.6	100	20
Thelandros sp.	90	1-17	3.3	64.6	5.7

The numerical and temporal occurrence of the nematodes is summarized in Table 2. It should be noted that of the 17 worms constituting the maximum infection by *Thelandros*, only one was an adult worm; the maximum number of adult *Thelandros* in any one host was five. Similarly, the heaviest *Oswaldocruzia* infection, 15 worms, consisted of immature individuals; the maximum number of adult worms in any one host was ten.

The monthly variation in the relative occurrence of young stages *versus* adult in both nematodes (Table 2) suggests that the parasites are eliminated from hosts sometime in the long period, late September to early June, when *A. hardii* exists subterraneously; the worms thus would be reacquired annually when the salamanders resumed living on the "surface" or near the surface. Table 2 shows that the majority of the worms are immature (100 per cent, in *Oswaldocruzia*) in samples taken in July. Additionally, all but one individual of those constituting the 20 per cent occurring as immature *Oswaldocruzia* in the period August to September were actually collected in early August. These were found in one salamander, and this constituted the heaviest infection for the period; crowding effects may have led to retardation of development of the worms.

If it is true that parasites are reacquired each spring—we assume that no temperature factors or immune reactions are delaying development of the worms, and no unusually long external ovic or free-living phase is a necessary part of their life-history—then the host-parasite data can be used as a basis for hypothesizing about the winter life of the salamander. During "surface" life the incidence of parasitism is high (90 per cent and 83 per cent: see Table 2), indicating that salamanders are readily invaded in times of activity. Salamanders examined in September were all parasitized and probably carried nematodes with them into their winter retreats. This part of their habitat should thus be contaminated with infective stages of both parasites. Yet the salamanders seem to become re-infected when the period of summer activity starts (note the high incidence of immature parasites in salamanders taken in July); therefore, the salamanders lose their worms in winter. This suggests that during their subterranean life salamanders are inactive, and avoid ingestion of infective stages of the parasites. A fairly complete hibernation such as we suppose they undergo has been reported by Szymanski (1914) for *Salamandra* on the basis of kymographic records of movement.

CHARACTERISTICS OF BREEDING SEX-RATIO

Tables 3 and 4 show the distribution of sexes for two subsections of our sample. The ratio of males to females in the total sample was nearly 1:1. There were differences in ratios between the three general localities: the two northerly sites had fewer females than males, when compared with the Cloudcroft samples. This is true for the samples of adults, but not for the juveniles, where in each instance the females predominated. We cannot absolutely explain these differences in ratios. Possibly the data on adults reflect different patterns of activity among the sexes so that adult females are simply not present in numbers where we looked for them. They could be located underground, in connection with "incubating" duties; if this is true it would account for the fact that so few egg-clusters have been found in logs.

TABLE 3.—SEX RATIOS IN ANEIDES HARDII, TOTAL SAMPLE

LOCALITY	Number of males	Number of females	Ratio of males to females
Capitan Mountains	35	31	100:87
Sierra Blanca	28	21	100:75
Sacramentos, 1958	23	20	100:121
Sacramentos, '56-'57	34	43	100:126

All	120	123	1:1

TABLE 4.—SEX RATIOS IN ANEIDES HARDII, ADULTS

LOCALITY	Number of males	Number of females	Ratio of males to females
Capitan Mountains	35	19	100:54
Sierra Blanca	22	7	100:32
Sacramentos, 1958	15	14	100:93
Sacramentos, '56-'57	22	16	100:73

AGE-RATIO

The data in Table 5 show adult salamanders to outnumber young at each collecting locality. This is probably not an accurate reflection of actual age composition in this species. Yet, we obtained the same general result in all three years of the study. We assume, therefore, that young were located where we could not catch many of them; probably they were underground. Sites of hatching and of the activities of early life would thus occur where we think the bulk of eggs are laid.

TABLE 5.—AGE RATIOS, ADULTS-JUVENILES

LOCALITY	Number of adults	Number of juveniles	Ratio of adults to juveniles
Capitan Mountains	57	15	100:26
Sierra Blanca	30	22	100:73
Sacramentos, 1958	42	30	100:71
Sacramentos, '56-'57	46	35	100:76
All	175	102	100:58

For purposes of this study we had only to age the individuals into adult and subadult classes. The criterion for adult status was breeding capability. A five-millimeter testis was the smallest size found in individuals that probably bred, and all of these were 40 mm. or more in snout-vent length. We arbitrarily considered individuals smaller than 40 mm. to be subadult. This probably does injustice to reality (females were treated the same way), but it should be noted that any error introduced in this way was almost certain to have increased the number of "subadults" in the samples. Thus, the hypothesis above based on age-ratios is not automatically invalid because of improper aging.

TIMING OF THE BREEDING SEASON

The time in which egg-clusters are deposited is a good rough index to events in the breeding cycle. We found four egg-clusters, one on July 14, 1957, and three on July 27, 1957; the only other eggs taken to date were found in late August (Lowe, 1950:94). Thus, courtship could occur in June, oviposition in July and August, and hatching from August to September. Actually, it is likely that the season is more restricted in time for any one year. Lowe's find was made in a year in which the summer rains were late, beginning in late July (Stebbins, 1951:137), whereas ours were made in a year having abundant and relatively early rainfall, beginning in late June. Microclimatic humidity is of extreme importance to both the salamanders and their food.

We suppose a great deal of breeding activity takes place underground; the chronology of events in such sites may bear no close relationship to those occurring at the surface, yet it is likely that a close parallel would be found. Breeding activities are ordinarily associated in time with greatest food abundance.

CLUTCH-SIZE

By clutch-size we refer to the number of eggs in laid clusters. We collected clutches of six, four, four and one; adding one more of three (Lowe, *op. cit.*) gives an average of 3.6 eggs per cluster; the average is 4.2 eggs if our clutch of one is discarded on the grounds it was incomplete.

For comparison we have listed (Table 6) clutch-sizes for some other plethodontids. It should be noted that these numbers refer only to eggs deposited in clusters, and not to large ovarian eggs. Thus, *Aneides hardii* has the lowest range in clutch-size of any North American plethodontid on record. It has been noted in other species that low clutch-size is correlated with low productivity, slow population turnover, and long average life-expectancy (Lack, 1954:103-105; Pitelka and Johnston, MS). If this is the case with this salamander, several other features in its environment and habits would tend to reinforce such population structure: the animals are exceedingly well-concealed (they were first described only 17 years ago [Taylor, 1941]), apparently have few natural enemies (one garter snake [*Thamnophis*] was collected within the habitat of the salamander in three years), apparently have few and benign parasites, and abundant and readily available food.

TABLE 6.—RANGES AND MEAN VALUES OF CLUTCH-SIZES IN SALAMANDERS OF THE NORTH AMERICAN PLETHODONTIDAE[1]

	Range	Mean
Desmognathus spp.[2]	11-40	20
Leurognathus marmorata	28	

Plethodon cinereus	3-13	9
Plethodon spp.	8-18	13
Ensatina eschscholtzii	12-14	13
Hemidactylium scutatum	30	
Batrachoseps spp.	7-74	
Aneides hardii[3]	1-6	3.6
Aneides spp.	7-19	13
Stereochilus marginatus		57
Pseudotriton ruber	72	
Manculus quadridigitatus	3-48	

[1] From Bishop (1947) and Stebbins (1951).

[2] Clusters of one and two occasionally found in *D. ochrophaeus*.

[3] This study, and from Lowe (1950).

EGGS AND "INCUBATION"

Our information concerning eggs essentially duplicates that already reported (see Stebbins, 1951). All egg clusters that we found were in small chambers within decomposing fir logs. In each instance the eggs were suspended from the roofs of the chambers. The clutch of six eggs was a compact mass, and the individual suspensory cables of the eggs were intertwined and fused with one another. The clutches of four eggs, although they too were compact clusters, had each suspensory pedicel distinct from the others. The surface of the eggs was lightly moist, but did not glisten with water, and each egg was completely free of the others. The outer coat of jelly of the fresh eggs measured about 6.4 by 5.7 mm. as they hung suspended; sizes were uniform and no egg was notably smaller or larger than the others.

We attempted to keep eggs artificially, but mold destroyed them after 12 days. We had difficulty keeping them wet without inundating them, for the climate at Las Cruces, New Mexico, where we kept the eggs, is exceedingly dry in summer. Until death, embryos were active and responsive to disturbances around them. This was at a time when the limb buds could not be detected and when the external gills were evident only under close scrutiny.

Two times we found adult female salamanders in the chambers with the egg clusters. The other two egg clutches seemingly had no attendant adult, but our method of going through a log was such that we could easily have alarmed any attendant animal well before we found the eggs, allowing time for the adult to move away from the eggs. We presume that incubation, so-called, in *A. hardii* is similar to that found in other plethodontids (see, for example, Gordon, 1952:683). Our findings on the conditions of the stomachs of these attendant adults have been outlined above ("Food and Foraging"). Our limited data suggest that only females are found in chambers with eggs.

SUMMARY

The montane relict plethodontid *Aneides hardii* was studied in the field and laboratory in 1956-1958. Food items detected in a small sample of stomachs are listed tabularly. Two roundworms were found to parasitize the guts of the salamanders; the parasitism looks to be benign. Subterranean winter inactivity is thought to be an integral part of the salamanders' lives, and is suggested in part by the life cycles of the worms. Summer activity appears to occur at the ground surface in logs and talus, and underground; the latter site is suggested by certain ratios obtained in the samples, showing adults to outnumber young and males to outnumber females. The season for egg deposition seems to be in July and August. Clutch-size is lower than for any other plethodontid on record. "Incubation" of eggs apparently parallels that characteristic of other plethodontids.

LITERATURE CITED

BISHOP, S. C.

1947. Handbook of salamanders. Ithaca, Comstock. xiv + 555 pp.

GORDON, R. E.

1952. A contribution to the life history and ecology of the plethodontid salamander *Aneides aeneus* (Cope and Packard). Amer. Midl. Nat., 47:666-701.

LACK, D.

1954. The natural regulation of animal numbers. Oxford, Clarendon, viii + 343 pp.

LOWE, C. H., JR.

1950. The systematic status of the salamander *Plethodon hardii*, with a discussion of biogeographic problems in *Aneides*. Copeia, 1950(2):92-99.

STEBBINS, R. C.

1951. Amphibians of western North America. Berkeley, Univ. Calif, xviii + 539 pp.

SZYMANSKI, J. S.

1914. Eine Methode zur Untersuchung der Ruhe- und Aktivitätsperioden bei Tieren. Arch. ges. Physiol., 158:343-385.

TAYLOR, E. H.

1941. A new plethodont salamander from New Mexico. Proc. Biol. Soc. Wash., 54:77-79.

Transmitted May 11, 1959.

27-9040

CPSIA information can be obtained
at www.ICGtesting.com
Printed in the USA
LVHW100100081222
734780LV00031B/1358

9 789356 707313